Mathematical Olymp

for

Elementary School 4

My First Book of Mathematical Olympiads – *Fourth Grade*

(Workbook)

My First Book of Mathematical Olympiads

Mathematical Olympiads *for* Elementary School

4

Fourth Grade

(Workbook)

Michael Angel C. G., Editor

Preface

The *Mathematical Olympiads for Elementary School* are open mathematical Olympiads for students from 1st to 4th grade of elementary school, and they have been held every year in the city of Moscow since 1996, their first editions taking place in the facilities of the Moscow State University - Maly Mekhmat. Although initially these Olympiads were conceived for students of a study circle of elementary school, then it was extended to students in general since 2005. Being the Technological University of Russia – MIREA its main headquarters today. Likewise, these Olympiads consist of two rounds, a qualifying round and a final round, both consisting of a written exam. The problems included in this book correspond to the final round of these Olympiads for the 4th grade of elementary school.

In this workbook has been compiled all the Olympiads held during the years 2011-2020 and is especially aimed at schoolchildren between 9 and 10 years old, with the aim that the students interested either in preparing for a math competition or simply in practicing entertaining problems to improve their math skills, challenge themselves to solve these interesting problems (recommended even to elementary school children in upper grades with little or no experience in Math Olympiads and who require comprehensive preparation before a competition); or it could even be used for a self-evaluation in this competition, trying the student to solve the greatest number of problems in each exam in a maximum time of 2 hours. It can also be useful for teachers, parents, and math study circles. The book has been carefully crafted so that the student can work on the same book without the need for additional sheets, what will allow the student to have an orderly record of the problems already solved.

Each exam includes a set of 8 problems from different school math topics. To be able to face these problems successfully, no greater knowledge is required than that covered in the school curriculum; however, many of these problems require an ingenious approach to be tackled successfully. Students are encouraged to keep trying to solve each problem as a personal challenge, as many times as necessary; and to parents who continue to support their children in their disciplined preparation. Once an answer is obtained, it can be checked against the answers given at the end of the book.

Sincerely,

The editor

Contents

Problems

Problems

Olympiad 2011

(XV Olympiad for Elementary School)

Problem 1. A family consists of four children, three fathers, a grandfather, and two grandchildren. How many people make up the family?

Problem 2. Gosha had a sheet of paper 4 cm wide and 10 cm long. If she folded it several times to make a rectangle 4 cm high and 1 cm wide, and then cut out the figure of a girl. (see picture) How many shapes did he get?

Problem 3. Cut the figure on the left into 4 equal parts.

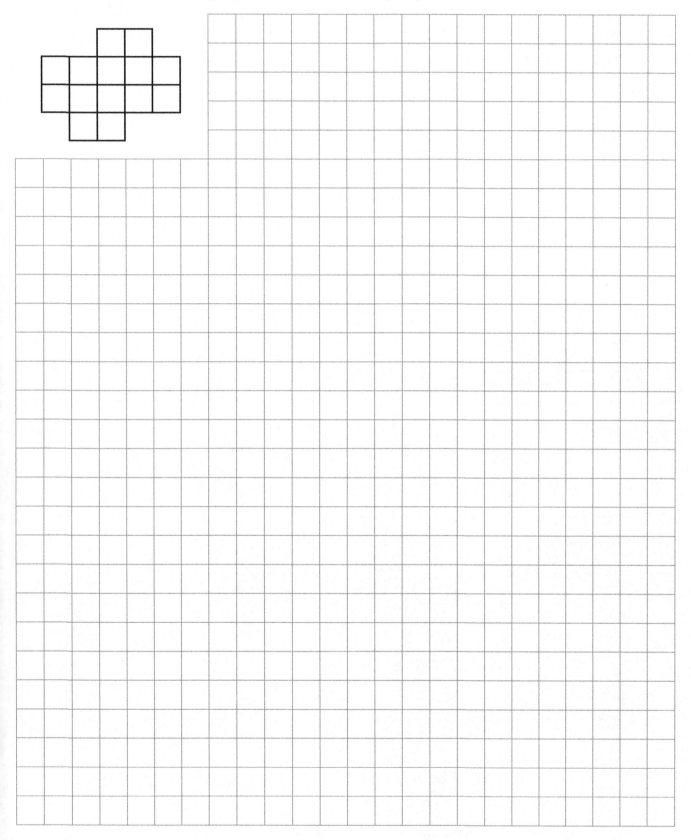

Problem 4. Kopatych weighs more than Losyash. Hedgehog and Losyash together weigh more than Nyusha and Kopatych together. But Kopatych and Losyash together weigh the same as Hedgehog and Nyusha together. Who weighs more and who weighs less?

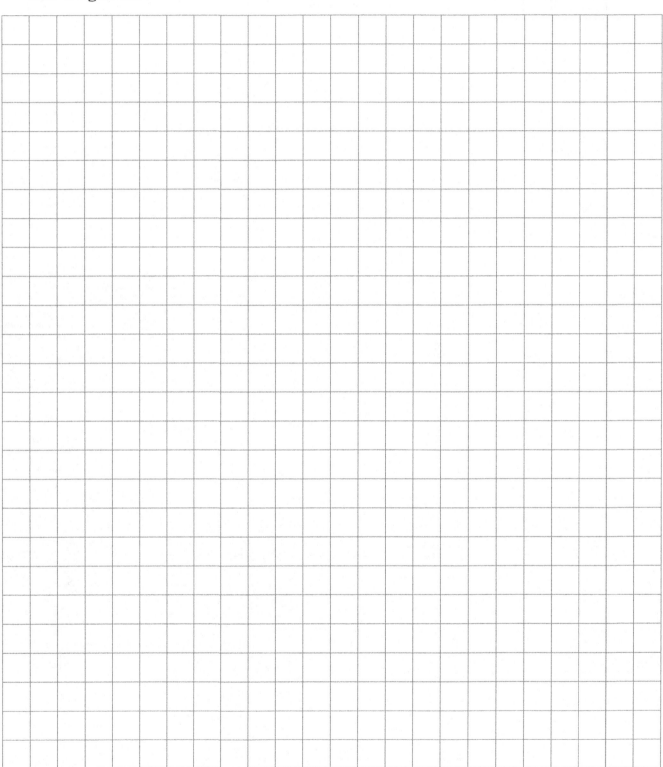

Problem 5. Ranger Stepanych roams the boundaries of his place in 5 hours. While the Ranger Mikhalych walks the boundaries of his place in 6 hours. When Mikhalych retired, his place was incorporated into the Stepanych place and now Stepanych spends 10 hours roaming the boundaries of the combined places. How long did it take him to travel the shared boundary of these places, if the speeds of the rangers are the same?

Problem 6. Dima has three large coils of rope, in blue, red and yellow colors. Cut 10 cm from the coils and tie three pieces into a 30 cm ring. How many different rings can he get?

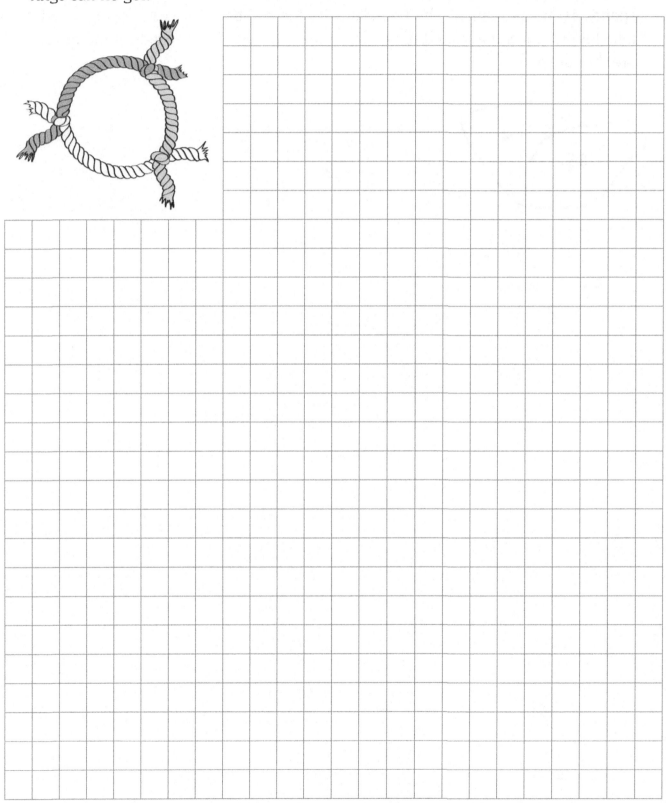

Problem 7. Vanya has 8 dominoes (see picture). She wants to arrange them in the shape of a 4 × 4 cell square so that the sum of points in all the rows and all the columns of the square is the same. A) What should this sum be equal to? B) How does Vanya need to arrange the dominoes?

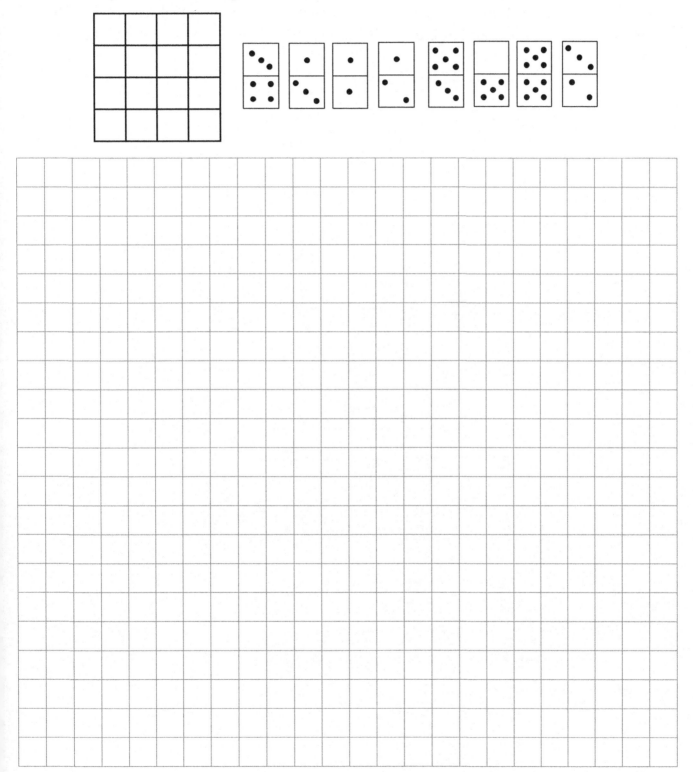

Problem 8. John Silver hid a treasure of gold and silver on three islands: Green Island, Amber Island, and Rocky Island. In one he hid the gold, in another he hid the silver, and in the remaining island he hid nothing. In the bay of each island, he posted signs. On the green island: "Gold on the rocky island". On the amber island: "There is no gold or silver here". And on the rocky island: "There is no silver either on the green island or on the amber island". Where is there definitely nothing if all the signs are not telling the truth?

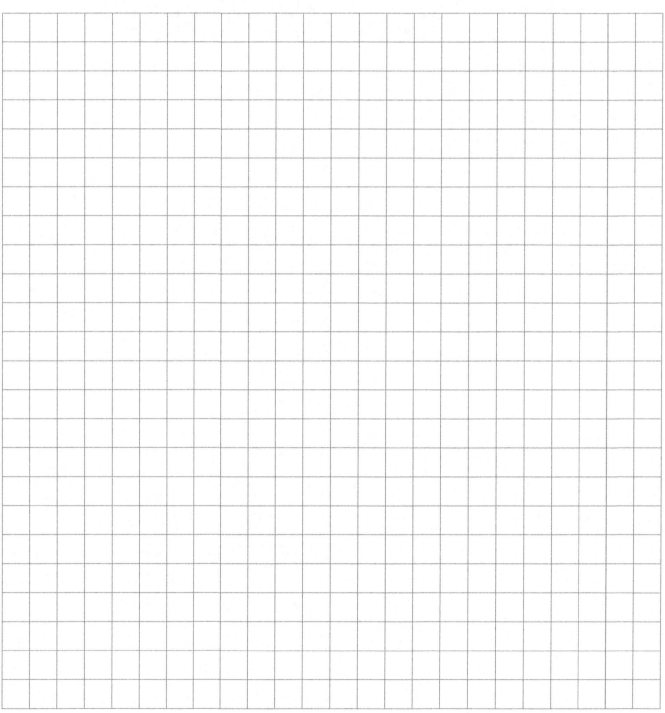

Olympiad 2012

(XVI Olympiad for Elementary School)

Problem 1. Swap the two-digit places to get the correct equality:

$$2012 = 1719 + 275$$

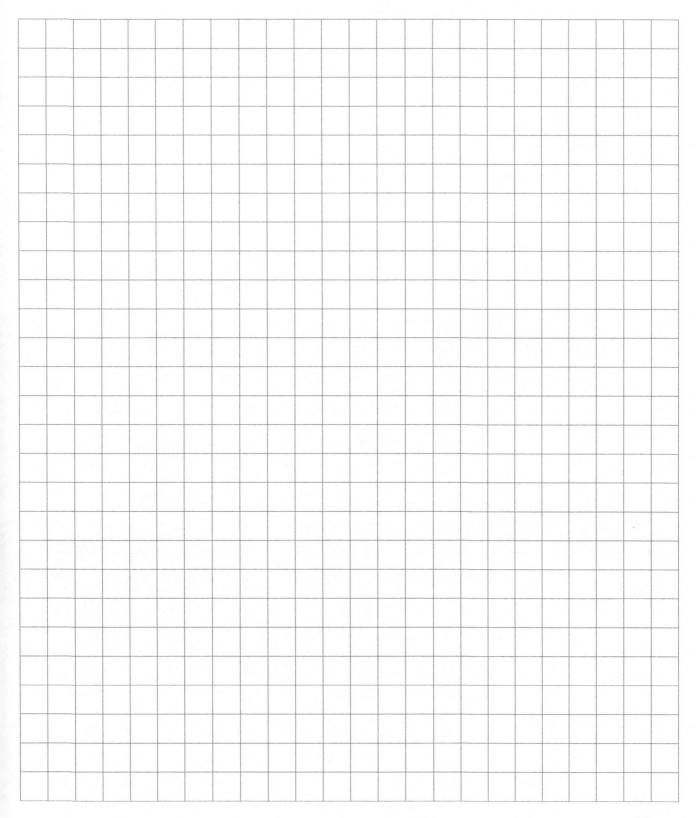

Problem 2. Petya ate a round cake on his birthday, which was sliced down the center. Each slice had a candle, and some slices also had a rose. Masha and Misha began to count the candles in a circle (each one started with a candle), but they both forgot where they started. Masha counted 6 candles and 2 roses, and Misha counted 19 candles and 3 roses. How old is Petya?

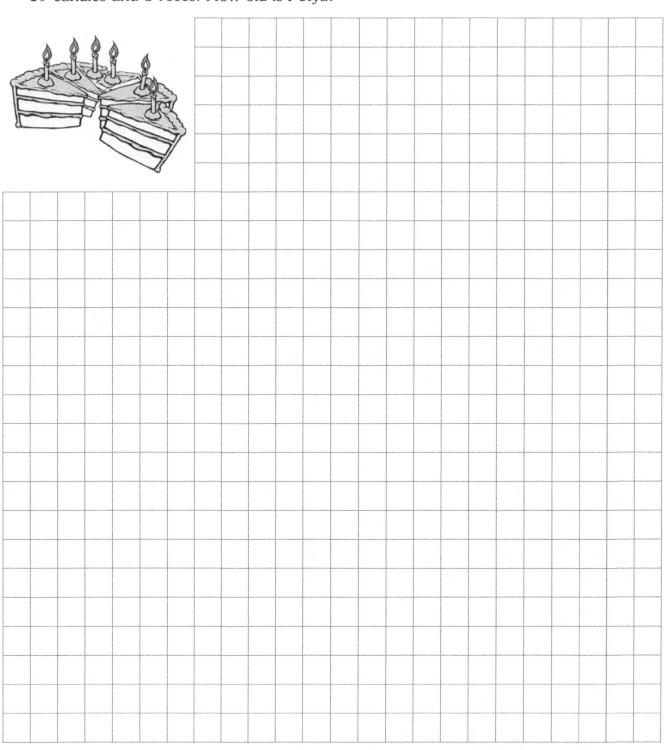

Problem 3. Cut the checkered figure into two identical pieces, each of which is an expansion of a $1 \times 1 \times 1$ cube.

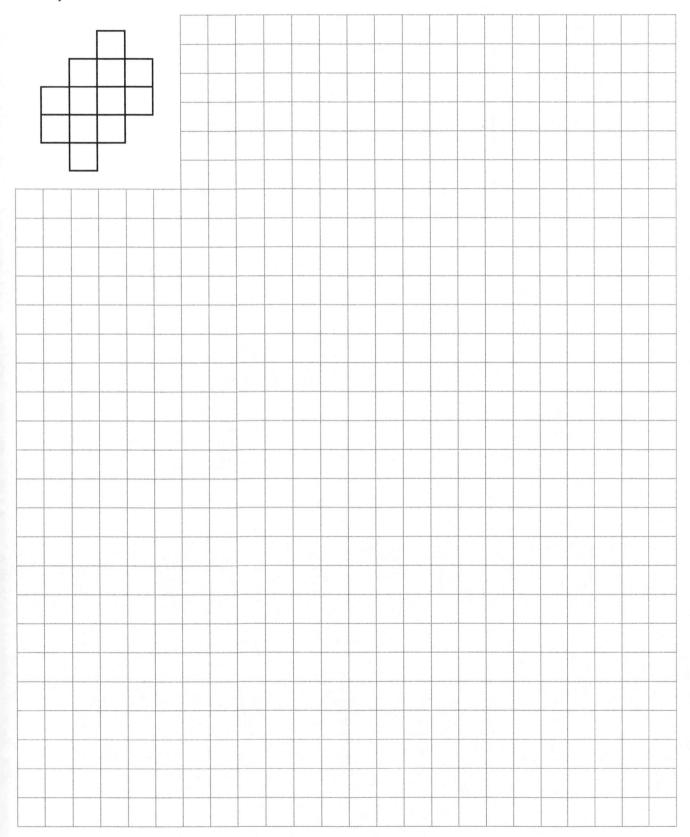

Problem 4. In the table below, place the numbers 1 through 7 so that each column and each row, as well as each highlighted figure, contains all seven numbers.

Problem 5. Nikita has a ruler with divisions in centimeters and millimeters. Likewise, Nikita discovered that there are exactly 80-millimeter divisions on the ruler. What is the distance between the first and the last division of Nikita's ruler?

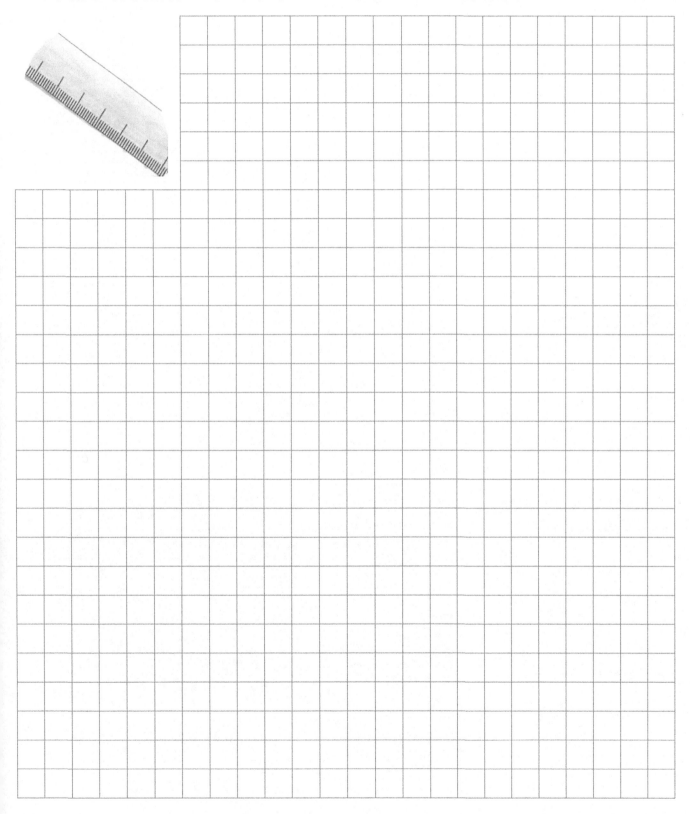

Problem 6. *Winnie the Pooh* has 11 large jars of honey and 10 small ones. A store sells boxes in which it can pack 5 large, 9 small, or 4 large and 3 small jars. How many boxes does Winnie have to buy to pack all of his jars of honey? (If he wants to buy as few boxes as possible).

Problem 7. Andrey, Borya and Vitya came to the Olympiad. One of them is a first grade student, another one is a second grade student, and the remaining one is a third grade student. It is known that the second grader solved one less problem than Andrey and Vitya solved two more problems than the third grader. Who solved more problems and how much more: Borya or a first grader?

Problem 8. Sasha has 2 gold coins, 3 silver and 4 bronze. One of them is false; If the false coin is silver, then it is lighter than real silver; and if the gold or bronze coin is false, then it is heavier than the real gold or bronze coin, respectively. Find the false coin after two weighings on a two-plate scale. (Note. Coins made of different metals may weigh differently, but real coins made of the same metal weigh the same.)

Olympiad 2013

(XVII Olympiad for Elementary School)

Problem 1. In the next sum, different letters represent different digits. It turned out OLIM + PI + ADA = 2013. Indicate which numbers could be in place of the letters.

Problem 2. I have two friends who hate dressing in the same clothes and shoes. Therefore, they always dress in such a way that everything is different about them. Find these guys among the seven guys shown.

SASHA   NIKITA   YASHA   VASYA   TOLYA   KOLYA   OLEG

Problem 3. The young physicist Ilya has two identical elastic bands. Likewise, he marked the midpoint of each of them and hung weights on their ends so that one elastic band became three times as long as the other. Ilya measured how far one mark is below the other. How many times is this distance less than the length of the longest elastic band?

Problem 4. On New Year's Eve, the hacker Kostya carried out 17 virus attacks on the Coca-Cola website, at regular intervals. The first attack began on December 31 at 9:54 p.m. and the last attack on January 1 at 11:30 a.m. What was the time interval between the attacks?

Problem 5. Anya, Borya, Vasya, Galya and Dasha decided to eat a chocolate bar. But this one fell to the ground, and when they picked it up, it turned out that it was broken into seven pieces, as shown in the figure. Borya ate the largest piece. Galya and Dasha ate the same amount of chocolate, but Galya ate three pieces and Dasha one. Vasya ate a seventh of the entire chocolate bar and Anya ate the rest. What piece of chocolate did Anya get?

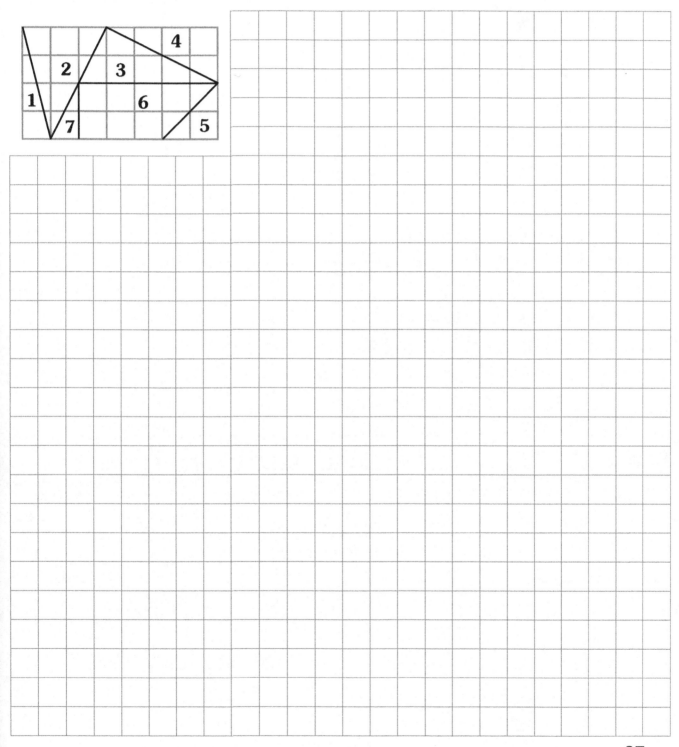

Problem 6. In the figure shown with matches, six triangles can be counted. Move four of these matches so that nine triangles are visible. Likewise, there should be no extra matches.

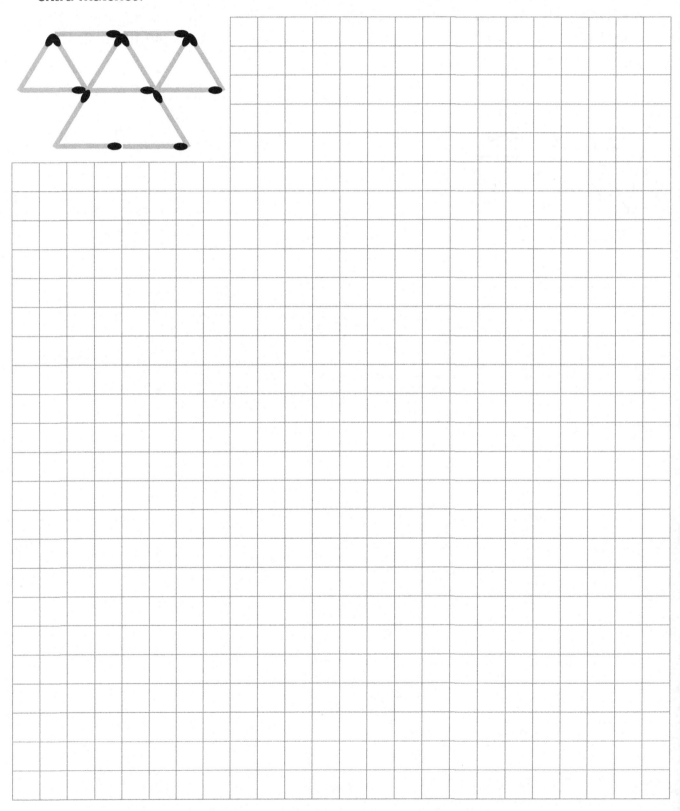

Problem 7. A tinsmith makes letter signs. He makes the same letters at the same time, and different letters, perhaps at different times. If he spent 50 minutes on two signs "DOM MODA" and "DINO", and made the sign "DOMINADO" in 35 minutes. How long does it take to make the sign "ANIDO"?

Problem 8. Once at a party, four friends were talking. Gloria said, "I always speak less than six words". Rico replied: "My statement has no more than eight words!". Alex said, "Gloria and Rico are now telling the truth". Marty added worried: "But today someone, Alex or Gloria, lied". Determine who lied and who told the truth.

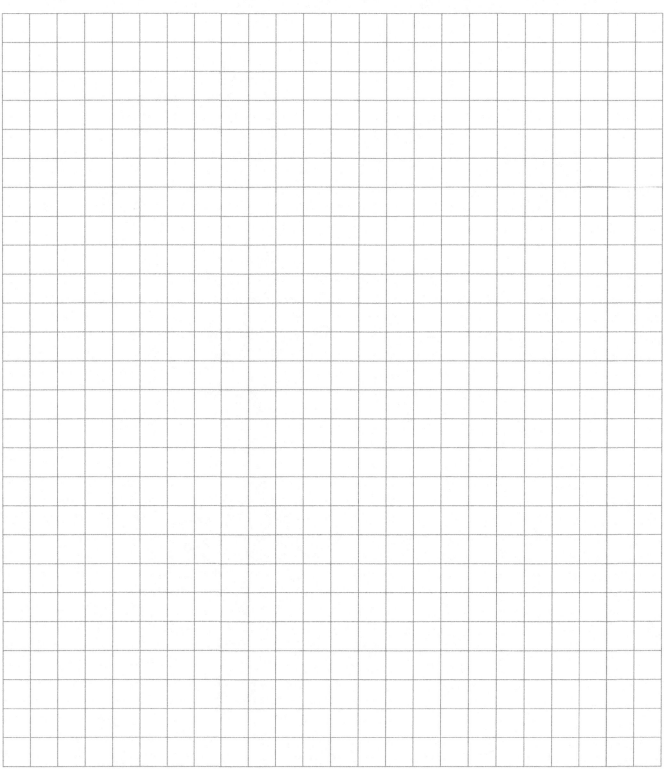

Olympiad 2014

(XVIII Olympiad for Elementary School)

Problem 1. A neighbor's father, son and grandson came to congratulate him on his birthday. His names are Anton Sergeevich, Andrei Borisovich and Sergei Nikitich. What is the name of the neighbor if he has only one son and no daughters?

Problem 2. The postman Pechkin left Prostokvashino and the policeman Svistulkin left Smetanino. They met on a kilometer post, on both sides of which were written the distances to Smetanino and Prostokvashino. Pechkin noticed that these are two different numbers, written with the same numbers, but in a different order. What is the smallest distance between Prostokvashino and Smetanino?

Problem 3. Sharik built a parallelepiped with 2 cm, 4 cm and 6 cm sides. Matroskin built a cube with a 3 cm side. Uncle Fyodor cut a rectangular hole in cardboard, into which Sharik's parallelepiped fits, but Matroskin's cube does not. What size hole could he cut? Just give 1 example.

Problem 4. In the figure made of matches, a small, a medium, and a large triangle are shown. From this figure, construct one in which there are exactly two small, two medium and two large triangles. There should be no additional matches, if each match is to participate in at least one triangle.

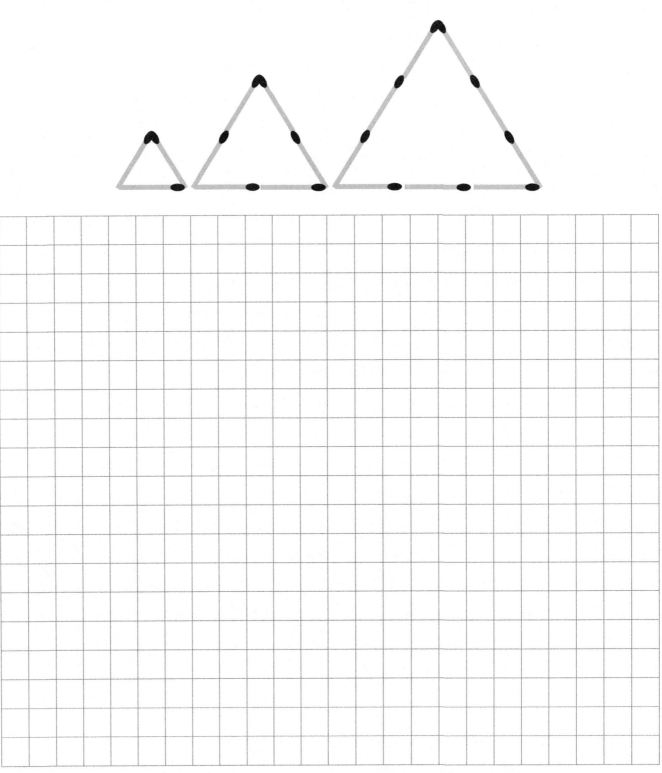

Problem 5. Examples of additions were written on a blackboard. Little Johnny replaced different digits with different letters. It turned that $O + N + E + F + O + R + O + N + E = 20$, and $S + I + X + F + O + R + S + I + X = 50$. What is the result of $O + N + E + F + O + R + S + I + X$?

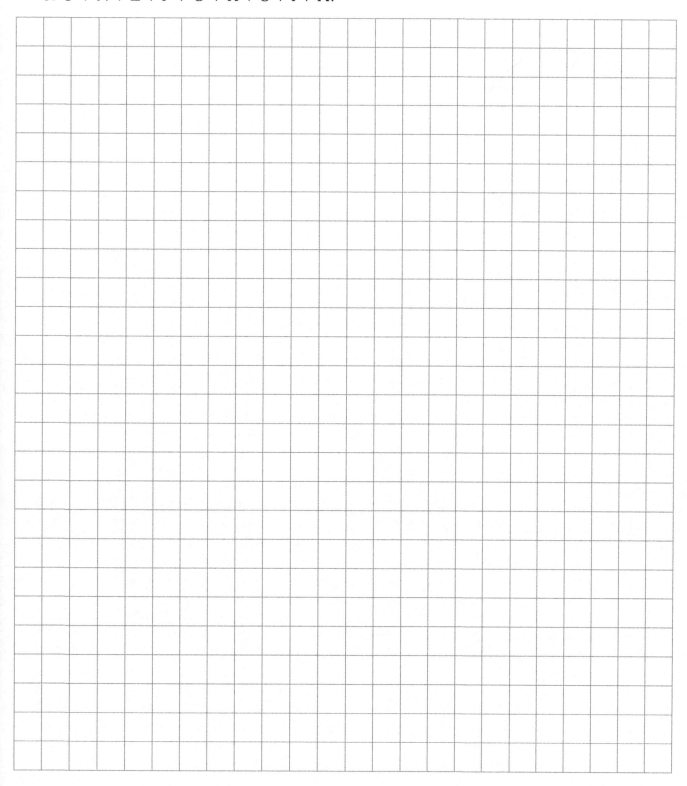

Problem 6. During some excavations in the territory of Ancient Rome, an unusual clock whose dial had 18 divisions was found, where Roman numerals were used for numbering (see figure). Unfortunately, the watch dial was divided into 5 parts. Nikita noticed that the sums of the numbers in each of the parts are equal. Show how the dial could have been broken.

Problem 7. Popeye "The Sailor" only eats spinach, and exactly once a day: either for breakfast, or for lunch or for dinner. It is known that if Popeye has breakfast on a certain day, the next day he will only have lunch. If he has lunch, the next day he will definitely not have breakfast. And if he dines one day, the next day he will necessarily have breakfast. Popeye had lunch on January 1, and during every day from January 1 to February 8, he ate breakfast as many times as he had dinner. What meal of the day did Popeye eat his spinach on February 8?

Problem 8. Brothers Avoska and Neboska only lie on their birthdays. Other days, they just tell the truth. Avoska once said: "Today is April 1. Tomorrow is your birthday". Neboska replied, "Today is your birthday. Tomorrow is April 1". When was Avoska born?

Olympiad 2015

(XIX Olympiad for Elementary School)

Problem 1. The Malinkins Masha and Vasya came to visit the Strawnichkins Petya and Sveta. And they brought them food: cookies, cake, chocolates and apples. The girls (Masha and Sveta) ate cookies and apples, and the Malinkins ate cookies and chocolates. They ate all the food; however, Masha does not like apples. What did each eat if everyone ate one of the foods?

Problem 2. Nikita wrote a puzzle about an inequality with two-digit numbers: ON > NO, where different letters represent different digits. How many solutions are there to this puzzle?

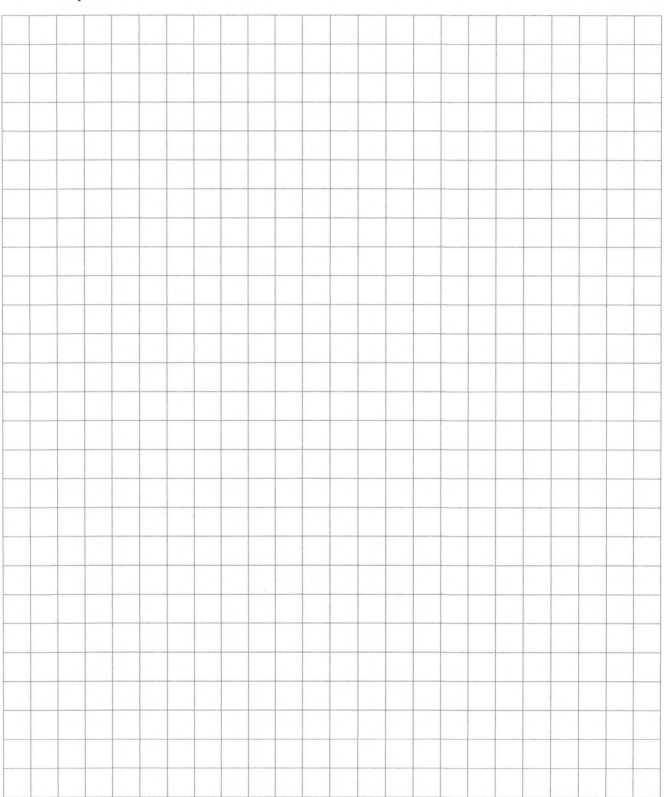

Problem 3. Yegor has chocolates and candies. If his mother gives him 10 more candies, the number of candies will be twice as many as the chocolates. Yegor wondered how many chocolates he should give Vitalik so that the remaining sweets also contain twice as many candies as there are chocolates. Help Yegor solve this problem.

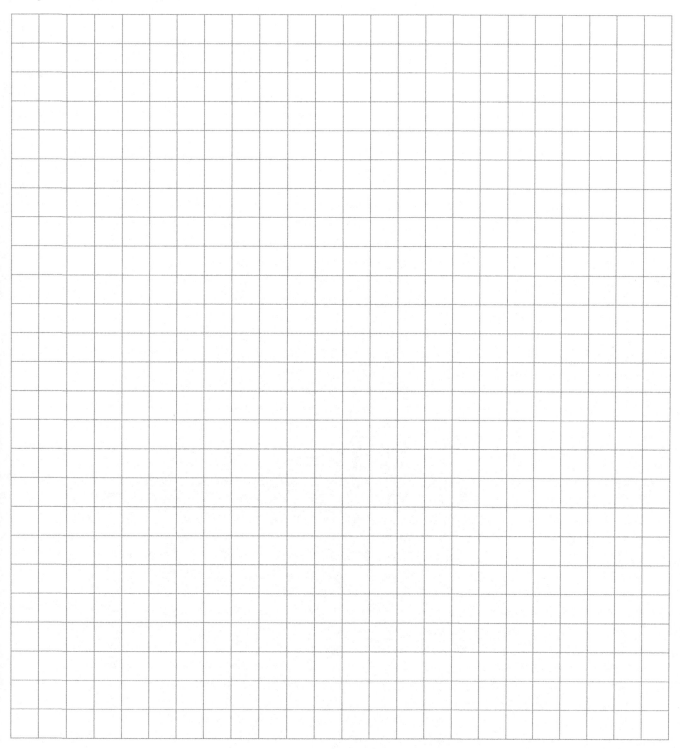

Problem 4. In 2015, Nikita will be so old that his age will be equal to the sum of the digits of her year of birth. What year was Nikita born? Find all the possibilities.

Problem 5. An arrangement with matches is shown in the figure below. Also, 3 rhombuses and 4 triangles can be seen. Move 4 matches so that only 1 rhombus and 6 triangles can be seen. (There should be no extra matches)

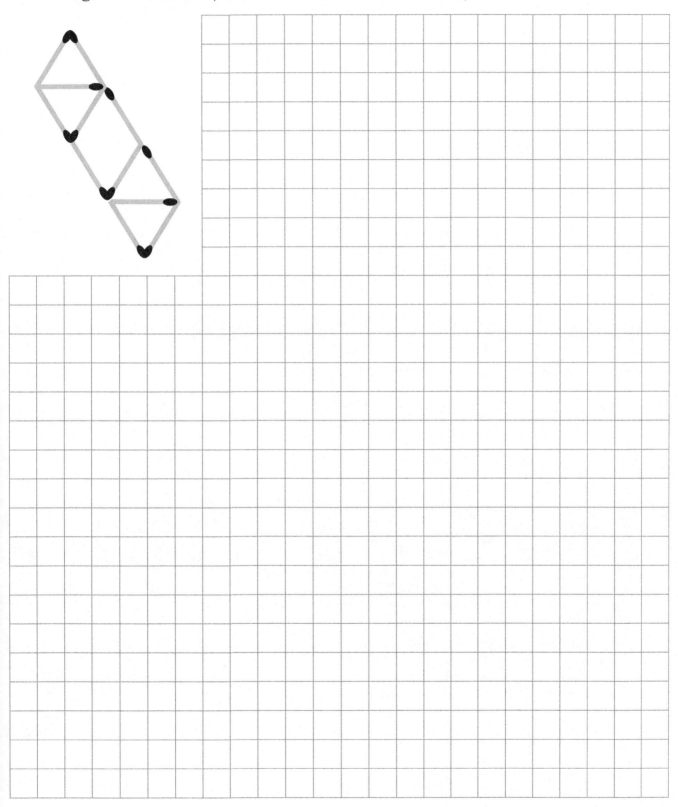

Problem 6. A bar of chocolate consists of 12 squares of black chocolate and 12 squares of white chocolate, as shown in the figure. Carlson wants to cut a 2 × 2 square so that there are equal parts of black and white chocolate. How many ways can he do it?

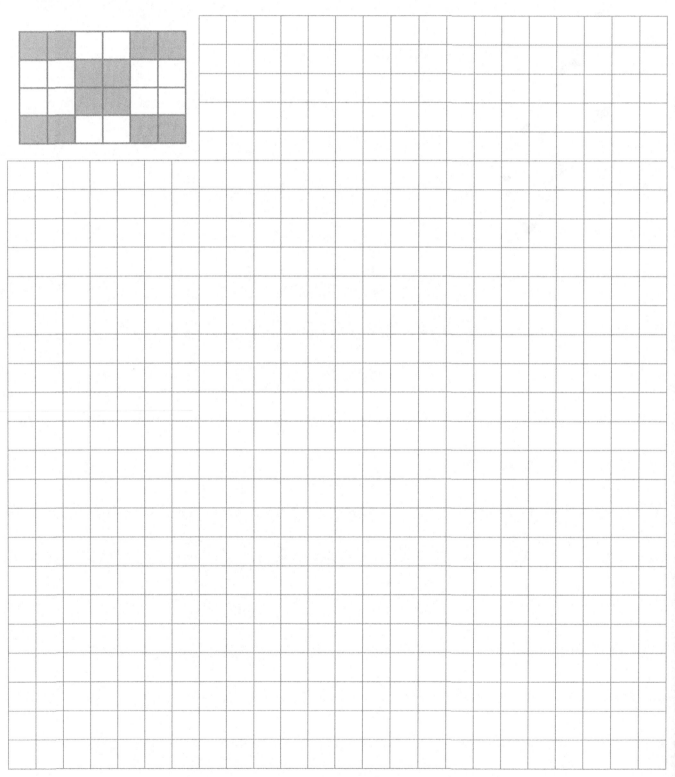

Problem 7. Some children observing a magician and his magic hat noticed that if he puts a handkerchief in his hat, after a minute he takes out a ball; if he puts a ball in his hat, after a minute he takes out a rabbit; if he takes something out of his hat, after a minute he waves his wand; if he waves his wand, in a minute he takes a handkerchief out of his pocket; if he takes a handkerchief out of his pocket, in a minute he puts it in his hat; if he takes a rabbit out of his hat, in a minute he takes a ball out of his pocket; if he takes a ball out of his pocket, in a minute he puts it in his hat. Now the magician took a ball out of his pocket, what will he do in 7 minutes?

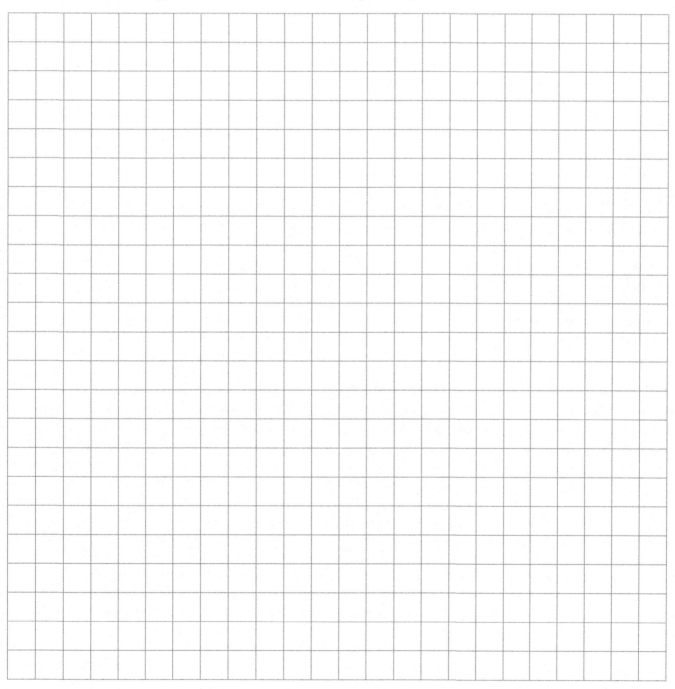

Problem 8. A family with three children lived on this island. One day they got together and said among themselves; Sasha: "I have two sisters". Zhenya: "And I also have two sisters". Valya: "And I have two brothers". How many boys and how many girls are there in this family? (Note: the names Sasha, Zhenya and Valya can be used by both boys and girls)

Olympiad 2016

(XX Olympiad for Elementary School)

Problem 1. Three brothers have a different number of coins. If the oldest gave 1 coin to the middle and the middle gave 3 coins to the youngest, then they would all have the same amount. How many coins must the oldest to the youngest of them give so that both have the same?

Problem 2. Place the numbers 1, 2, 3, 4 in the cells so that all 4 numbers are present in each row and in each column, and the indicated inequalities are satisfied.

Problem 3. Cut a 6 × 6 square board into two equal 24-sided polygons.

Problem 4. Ekaterina Mikhailovna (EM) had a collection of 10 notebooks with covers of different colors: Red (R), White (W), Black (K), Yellow (Y), Blue (B), Purple (P), Orange (O), Light Blue (L), fuchsia (F) and Green (G). Ten children formed a circle and EM began to distribute the notebooks to one in three children, counting in a circle and skipping those to whom she had already given. In what order were the notebooks, if Yegor received a yellow notebook, and he was the third to receive the notebook from her? The figure shows which notebooks all received in the end.

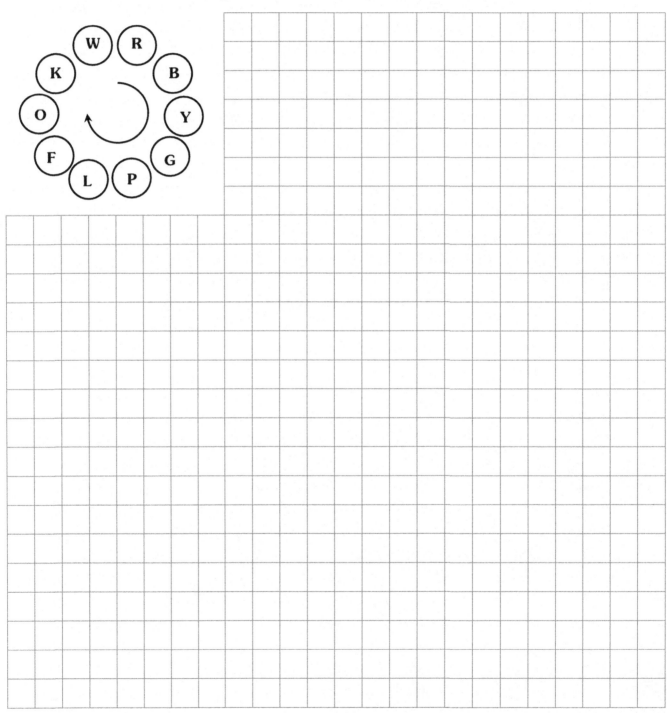

Problem 5. Find at least one solution to the puzzle. (If different letters correspond to different digits)

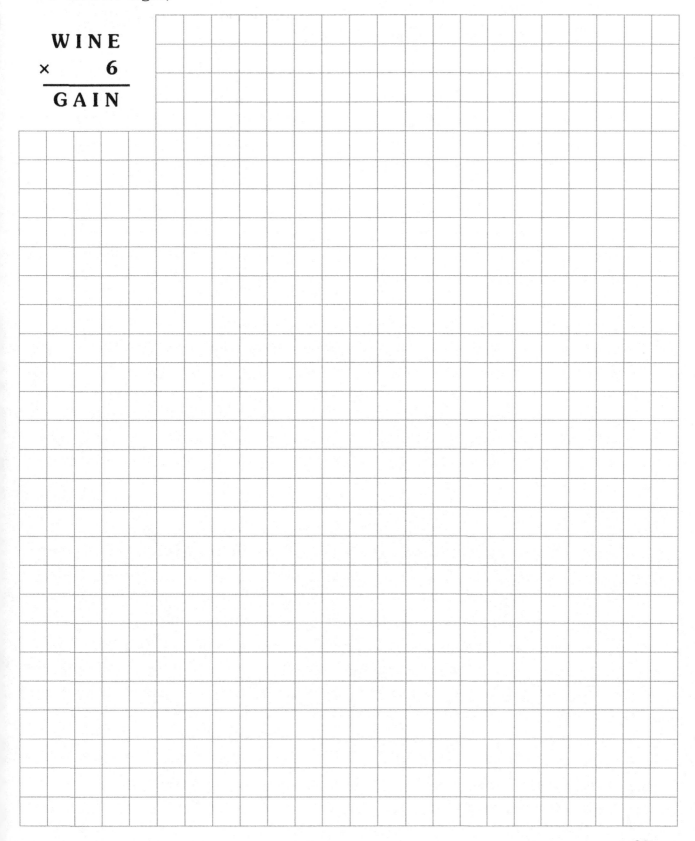

```
  W I N E
×     6
─────────
G A I N
```

Problem 6. Masha and Sasha brought an identical packet of cookies to school and agreed to eat 2 or 3 of them at each break. If at the end of the fourth lesson, Sasha had only one cookie left; and by the sixth lesson, Masha had already run out of cookies. How many cookies were in the packet?

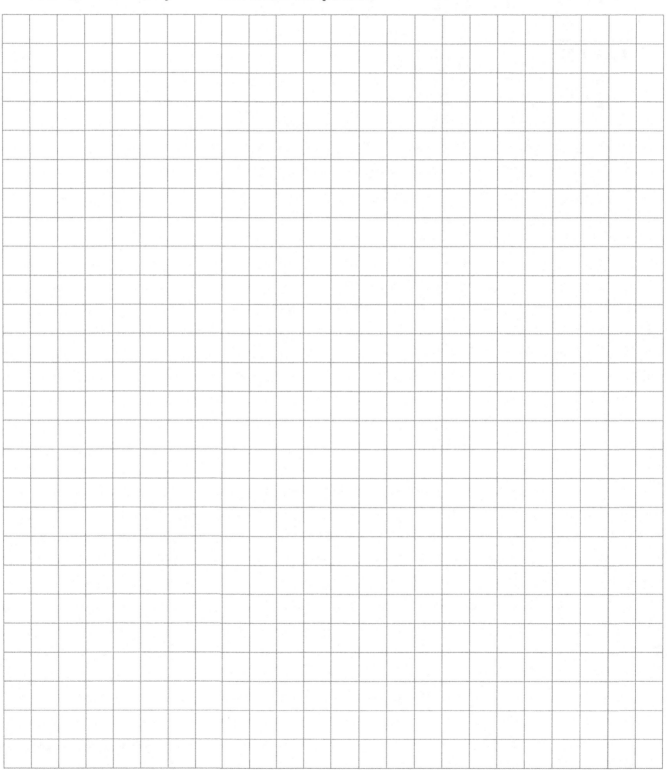

Problem 7. Once, a prince met with three witches to ask them about their fate and they answered him. Arta: The prince will have a lazy wife, and will defeat more than 100 dragons. Binah: No, no, the prince will defeat less than 100 dragons, but the wife will be a hard worker. Veda: No, the wife, alas, will be lazy, but the prince will defeat at least one dragon. What awaits the prince if it is known that one of them always lies, another always tells the truth, and the other one tells the truth first and then lies?

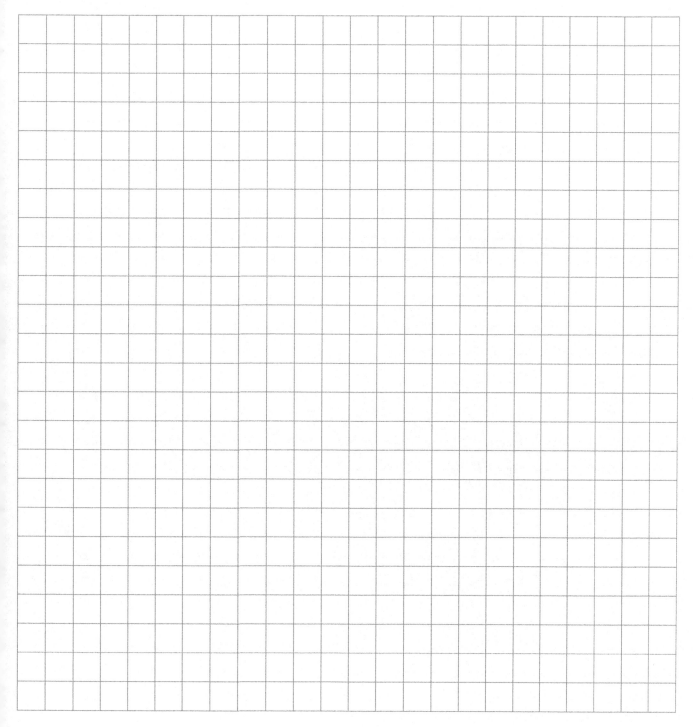

Problem 8. There are 9 coins in a row, it is known that among them there are exactly three that are false, and they are always together. All counterfeit coins weigh the same and are lighter than real coins. And all real coins weigh the same. How to find the three counterfeit coins in 2 weighings on a two-plate scale?

Olympiad 2017

(XXI Olympiad for Elementary School)

Problem 1. There are 9 cards with numbers from 1 to 9. Arrange them in a row so that there are not three consecutive cards with numbers in ascending or descending order.

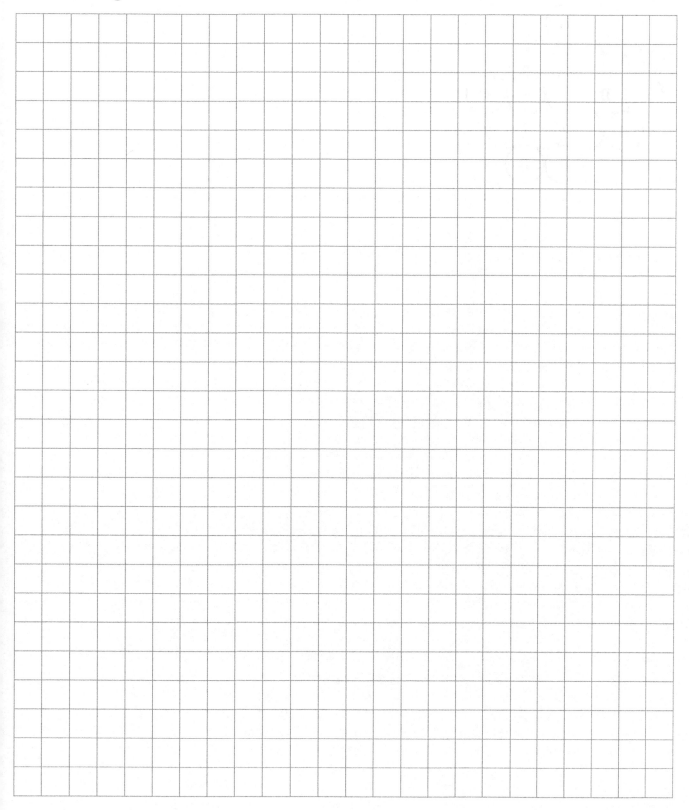

Problem 2. Place the letters A, B and C in the circles so that there are no equilateral triangles with three identical letters at their vertices.

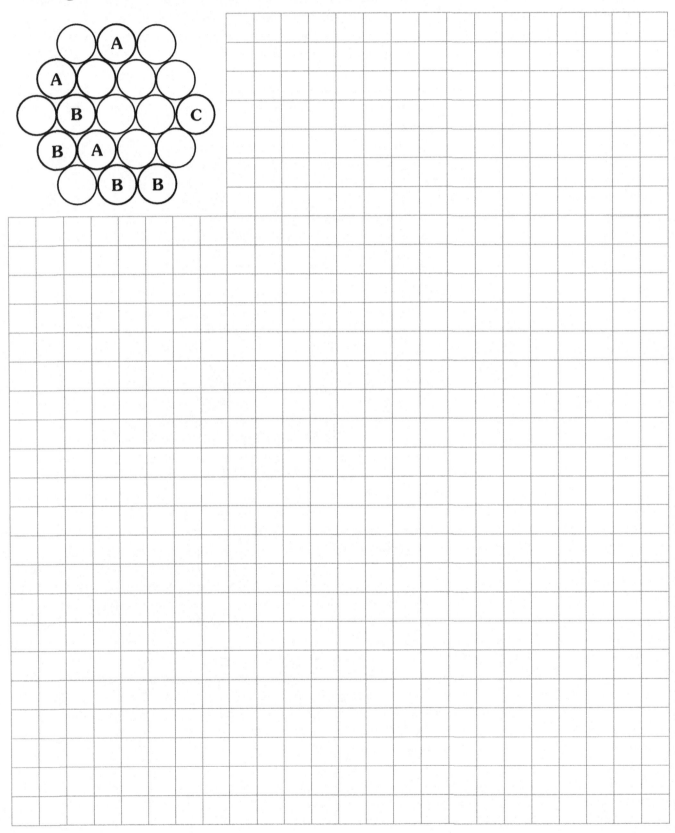

Problem 3. The city has 8 train stations: Alf, Beta, Hammilton, Dhelta, Lambda, Epsilon, Eks and Zeta. It is known that a train runs directly between two stations if the number of letters in the names of these stations has different parities. Fedya wants to travel as long as possible without visiting any station twice, so that the name of each subsequent station is longer than the previous one. How long (stations) will this journey take? Explain your answer.

Problem 4. The planet Zhelezyaka makes one revolution around its axis in 5 hours Zhelezyaka. And the planet Kamenyuk makes one revolution around its axis in 6 hours Kamenyuk. A spaceship travels from the planet Zhelezyaka to the planet Kamenyuk taking 20 hours Zhelezyaka, and returns in 25 hours Kamenyuk. Which planet rotates on its axis the fastest? Explain your answer.

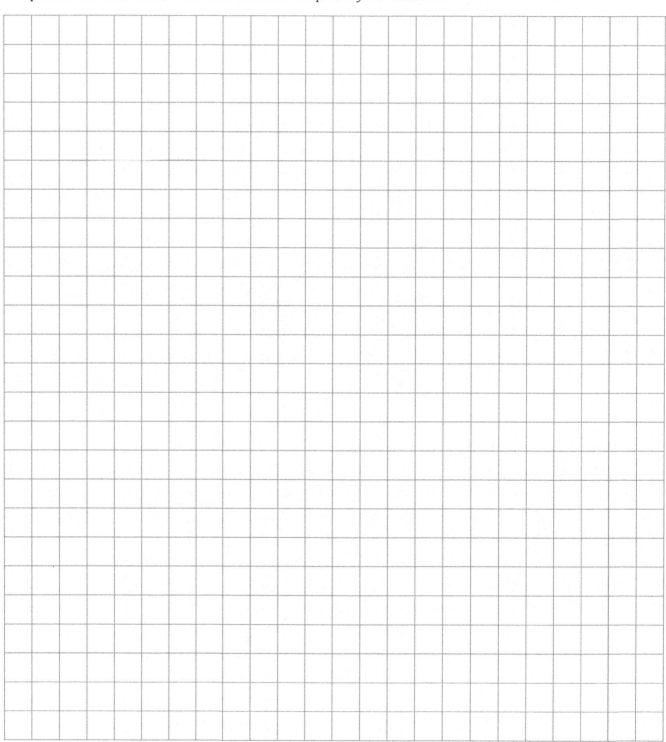

Problem 5. It was decided to number the houses along the only street in the City of Flowers, for which plates with numbers were used. It turned out that there were 12 more plates with the number 1 than plates with the number 0. What is the least number of houses on this street? Explain your answer.

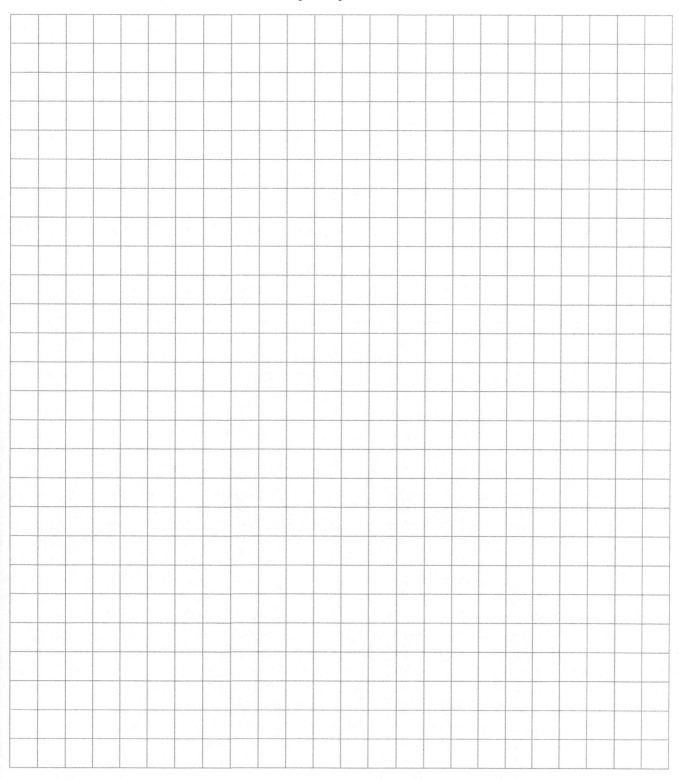

Problem 6. Cut the tree in the figure into 13 pieces with just three straight cuts.

Problem 7. Valya, Sasha, Zhenya and Slava played at recess. One of them broke a window, and each of them said; Valya: "One of the men broke it"; Sasha: "It was Slava!"; Zhenya: "There are more men among us"; Slava: "Valya and I are women!". It turned out that all the girls had lied and all the boys had told the truth. Who broke the window? (If all names can be both boys and girls) Explain your answer.

Problem 8. It is required to put an odd number of candies in boxes of 46 units, but only 43 boxes were filled. Then, it was tried to put them in boxes of 43 units, but 47 boxes were filled and there was also something left. Will it be possible to arrange the candies in exactly 17 boxes? Explain your answer.

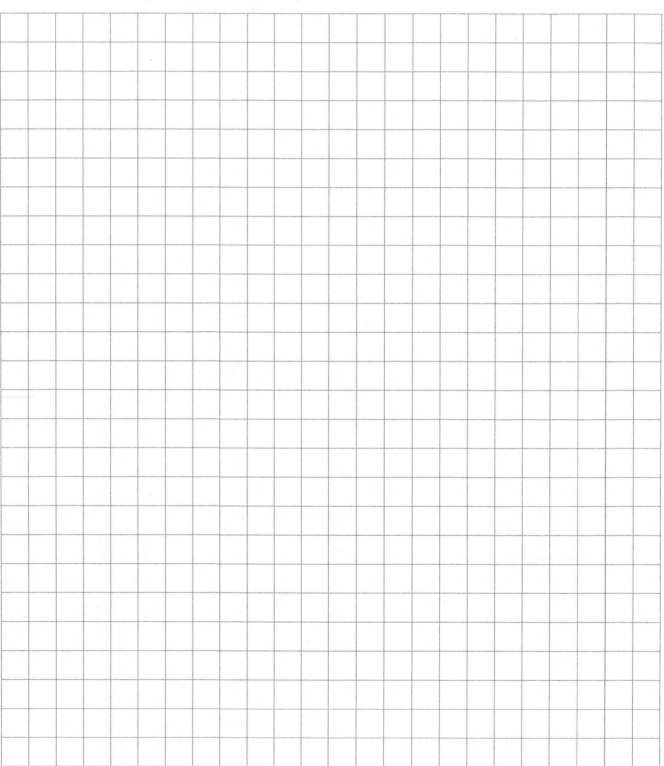

Olympiad 2018

(XXII Olympiad for Elementary School)

Problem 1. A cake costs the same as two turnovers and three turnovers cost as much as two chocolates. What is more expensive, two cakes or three chocolates?

Mathematical Olympiads for Elementary School – *Fourth Grade*

Problem 2. We can make numbers with the dominoes. For example,  represents the number 4203. Perform the correct operation of adding two four-digit numbers with the following dominoes:

84

Problem 3. A $4 \times 4 \times 4$ cm^3 cube was cut into $1 \times 1 \times 1$ cm^3 cubes. Then, from all these cubes, a rectangular frame was made for a photograph 1 cube thick (similar to the one in the figure). It turned out that the area of the photo that fits into the frame is 216 cm^2. Determine the dimensions of the photo.

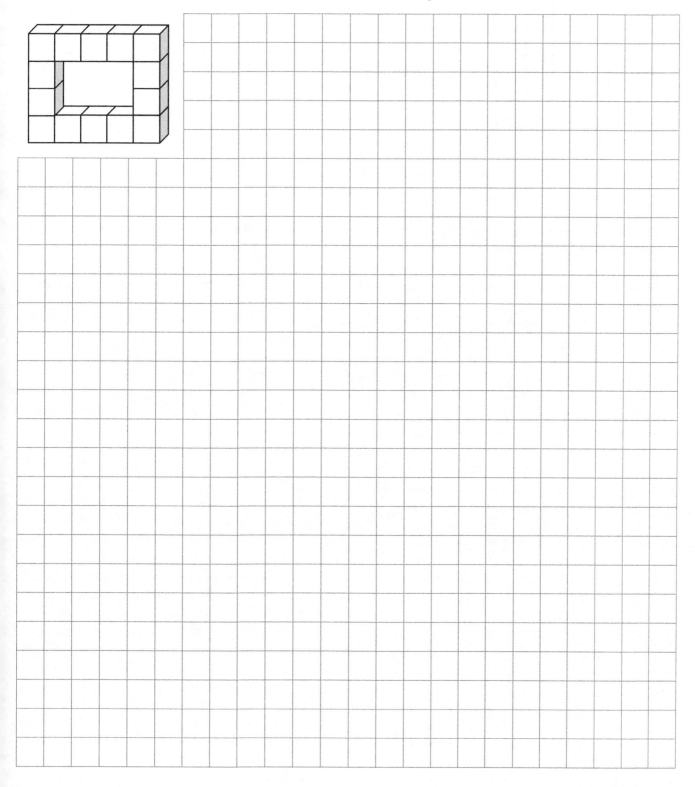

Problem 4. Cut the shape along the grid lines into 3 equal parts. (The cut parts may be rotated)

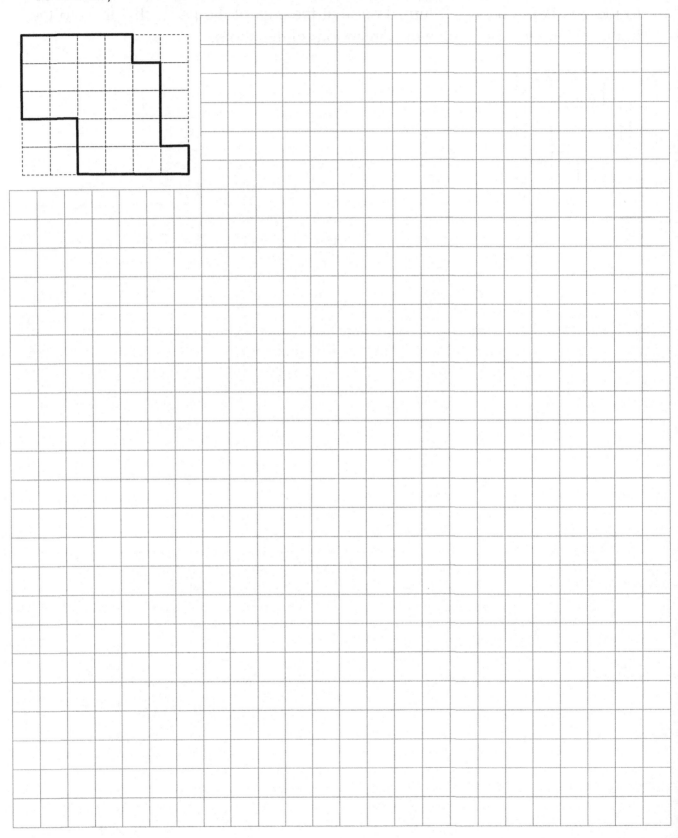

Problem 5. On Raduzhnaya Street, all the houses are located in a row. Each of the houses is painted in one of 5 different colors. It turned out that for two of these colors, it can be found neighboring houses painted in these two colors. What is the smallest number of houses on Raduzhnaya Street?

Problem 6. Some robots are trapped in a square-shaped trap made up of 36 square cells surrounded by columns (see figure). Each robot fires a laser simultaneously in 8 directions: the 4-sided and the diagonals. Robots are not in the line of fire of others. Likewise, they all fired at the same time and all the shots hit the columns. If the diagram shows the number of hits to each column. How many robots were there and where could they be?

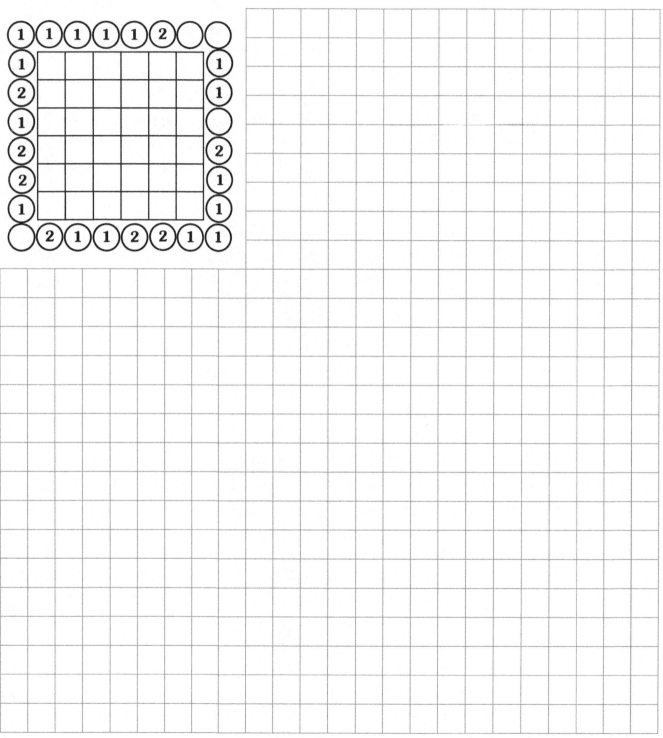

Problem 7. A mischievous monkey, a donkey, a goat and a black bear began to play in a quartet. They have violins (V), flutes (F), drums (D) and guitars (G). First, the bear and the monkey played guitars, the goat played the drums, and the donkey played the flute. Then they began to switch instruments. The bear every 15 minutes in this order: G-V-F-D-G-...., the monkey in the same order, but every 10 minutes. The donkey began to play in the order: F-G-V-D-F-... every 20 minutes. And the goat in the order D-F-V-G-D-... every 12 minutes. If they rehearsed for 2 hours and it is known that they only managed to play harmoniously when everyone had different instruments. How long did they manage to play harmoniously in 2 hours of rehearsal?

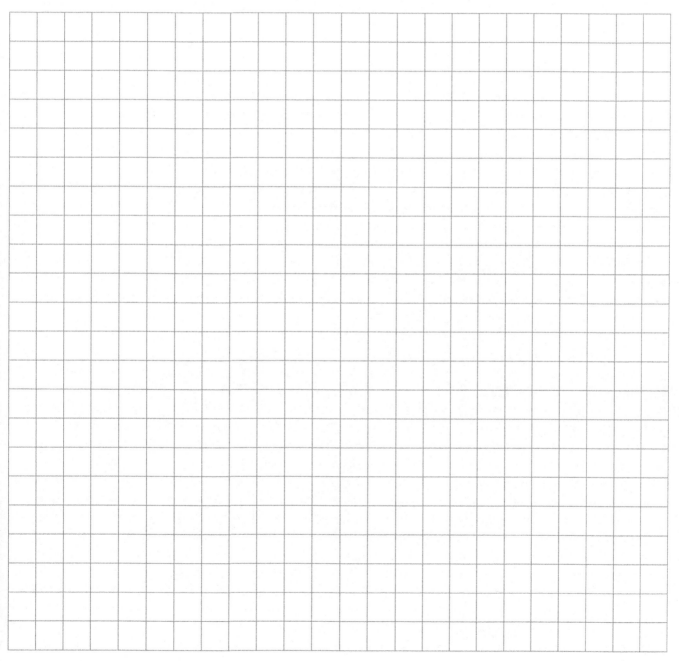

Problem 8. Five inhabitants of the island of gentlemen and liars were placed one after another. The last one (the fifth) said: "There are 4 liars in front of me". The fourth: "There are 3 liars in front of me". The third: "There are 2 liars in front of me". The second: "There is a liar in front of me". And the first was silent. How many of them are really liars? (If gentlemen always tell the truth and liars always lie)

Olympiad 2019

(XXIII Olympiad for Elementary School)

Problem 1. A boy removed the ornaments from a Christmas tree in a week. On Monday he removed 1 ornament and then each day he removed a number of ornaments equal to what he had removed on all the previous days together. If on Sunday he removed the last ornaments. How many ornaments were on the tree?

Problem 2. A boa (B), a monkey (M), an elephant (E), and a parrot (P) were weighed. The monkey wrote: Boa = 48 P, Elephant = 12 M, Monkey = 3 P, Boa = 4 M, Elephant = 36 P. It turned out that the monkey confused all the numbers, that is, the numbers were really the same, but they were all in different places (however, all the letters are written correctly). How many parrots do the boa, elephant, and monkey really weigh?

Problem 3. Put all the numbers from 0 to 9 in the cells to get the correct equality:

$$\boxed{} + \boxed{} \cdot \boxed{} - \boxed{} : \boxed{} = \boxed{}$$

Problem 4. A Christmas tree is drawn on graph paper. Cut it into 4 parts and form a square with them.

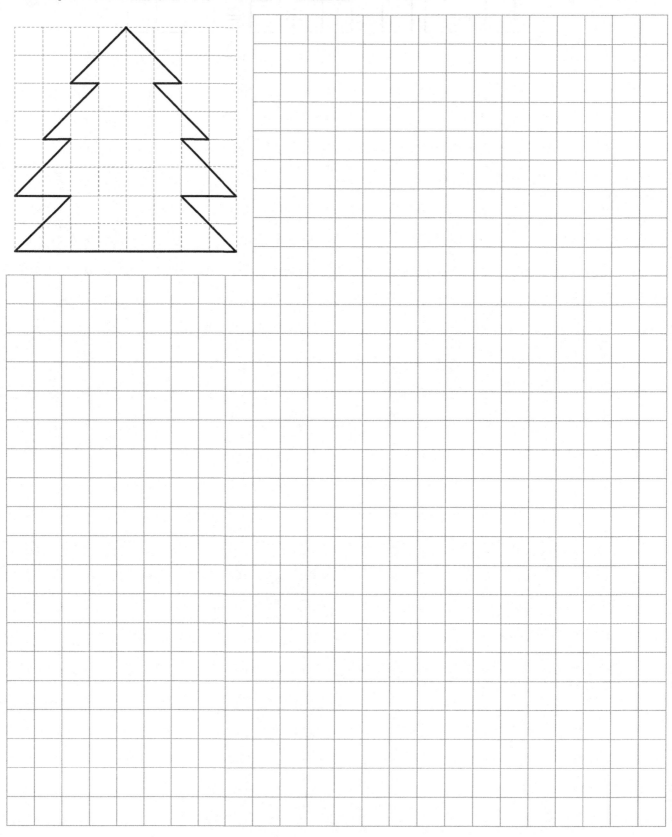

Problem 5. A group of children received 4 cards. Each card had one of the syllables PA, NA or MA written on it. It turned out that 13 children can form the word MAMA with their cards, 15 children can form the word PAPA and 17 children can form the word NANA. Likewise, 45 children can form the word PANAMA. How many children were there?

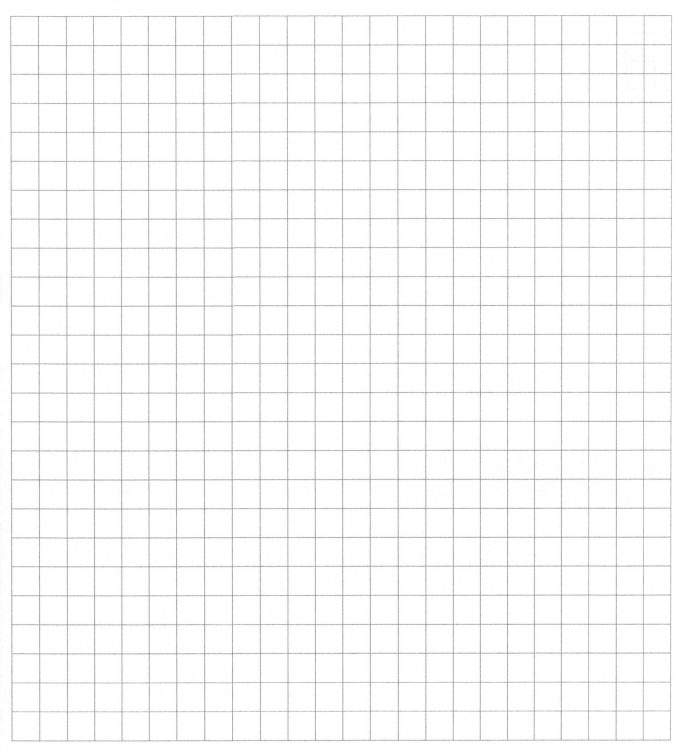

Problem 6. A rectangle was made from square cards (an example of a 6-card rectangle is shown in the figure). Then one side was reduced to its half and the other to its third part. If in the end 65 cards were removed. How many squares with a side of 4 cards can be formed from the original rectangle, without moving the cards? (in the figure it can be formed 2 squares with one side of 2 cards)

Problem 7. In the kingdom of the Full Moon, there are 9 cities as shown in the figure. The king wants to build direct roads between some cities so that they do not intersect outside the cities and there are exactly 4 roads leading out of each city. How could he do that?

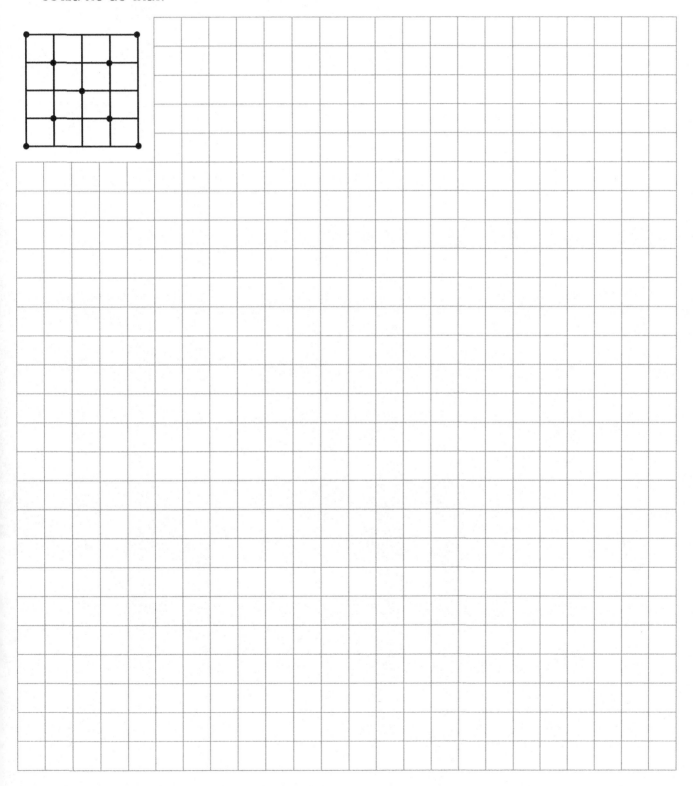

Problem 8. If a worm tells the truth, it turns green. And if it lies, it turns red. Once, two worms met. The first said: "We are both red". And then the second said: "If we had been silent, we would both be red now". Will the worms have the same color after this statement?

Olympiad 2020

(XXIV Olympiad for Elementary School)

Problem 1. Znayka multiplied two numbers and wrote down the resulting operation in encrypted form as: "MAKXIMUM". The signs for "multiplication", "equal", and each digit are represented by a letter. Likewise, different letters represent different signs or digits. What equality could Znayka have encrypted? Find at least one solution.

Problem 2. Vova swims a distance of 100 *m* in a 50 *m* long pool in 90 seconds, and in a 25 *m* long pool in 2 minutes. How long will it take for Vova to swim this distance in a 100 *m* long pool? Vova performs all the same actions and at the same speed.

Problem 3. If the smallest three-digit number that is not divisible by 4 is added the largest three-digit number divisible by 4. What is the result?

Problem 4. Kostya made a figure of three hexagons and wrote down numbers at all vertices, as in the left figure (he also wrote down a number at the central vertex; it is not known which one). Then Kostya increased the numbers at the vertices of one hexagon by a same number, then he increased the numbers at the vertices of the second hexagon also by a same number (possibly another), and then he did the same with the third hexagon. The right figure shows some of the numbers that were obtained. By how much has the number at the central vertex increased?

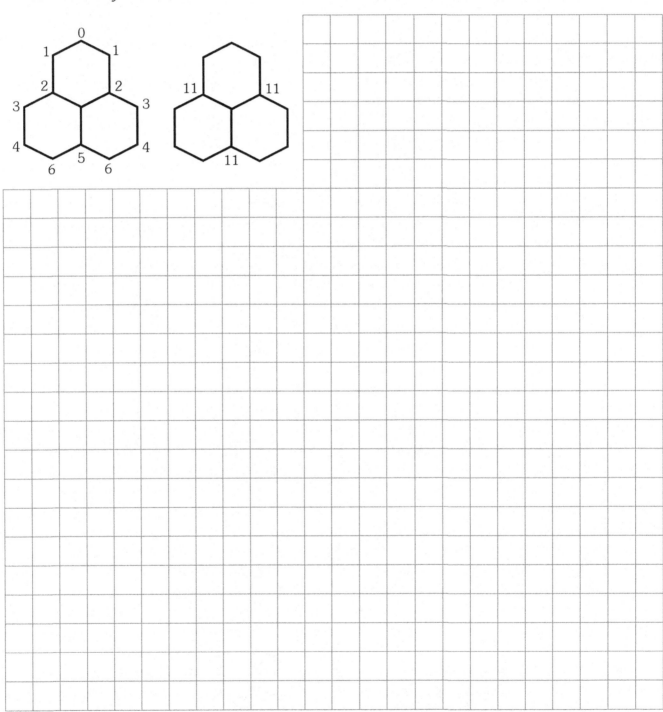

Problem 5. A park is divided into triangular sectors. If a flashlight is placed in one of the sectors, it will illuminate in three directions, as in the figure. Light up the whole park by placing 3 flashlights.

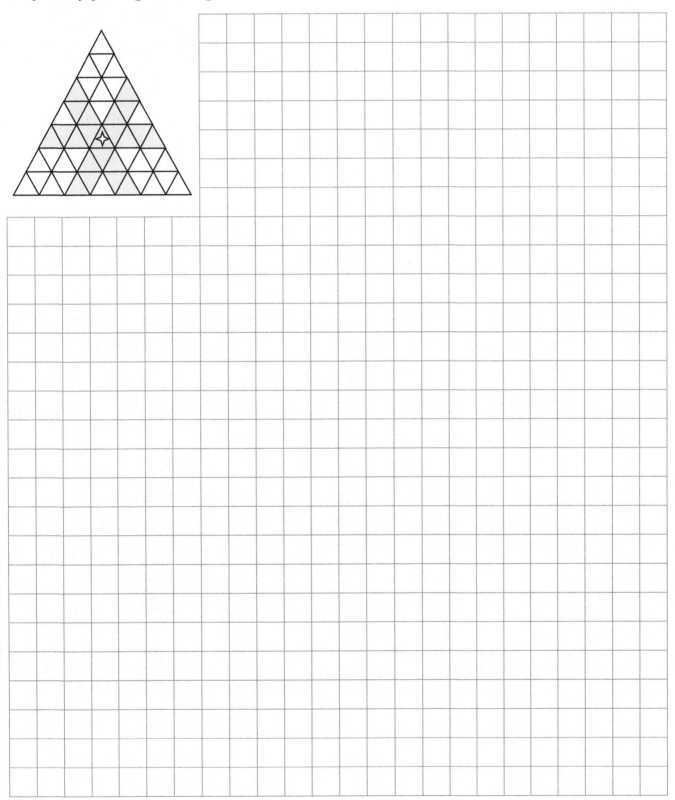

Problem 6. Pinocchio, Pierrot and Artemon were playing with snowballs. Pinocchio threw 20 snowballs, Pierrot - 14 and Artemon - 8. It is known that all of Pierrot's snowballs flew past. Artemon threw a snowball only in response to a snowball that fell on him, and exactly half of Pinocchio's snowballs hit their target. How many snowballs hit Pierrot?

Problem 7. From home to school, Klim has three intersections with traffic lights. From the first to the second traffic light, Klim takes 2 minutes and from the second to the third traffic light also 2 minutes. Klim knows that at each traffic light, the yellow light is on for 1 minute, the green and the red for the same time. Likewise, he takes 1 minute, at the first traffic light, at the second - 2 minutes, at the third - 3 minutes. Once, Klim saw through the window that at 8:00 a.m. at all traffic lights the green light was turned on simultaneously. At what time must he be at the first traffic light to get to school without stopping at them? (Klim crosses the street in 5 seconds)

Problem 8. Sasha, Kolya, Masha and Olya live on different floors of a five-story building. Once Sasha said: "I live above all the others!", Masha: "And I am in the middle!", Kolya: "I live above Masha and below Olya". And Olya added: "Kolya told a lie. Between Kolya and me there is an apartment where neither of us lives". It turned out that those who lived on the odd floors had lied and those who lived on the even floors had told the truth. Where does each one live?

Answers

Olympiad 2011

1. 5 people.

2. 5 figures.

3. An example is shown as follows:

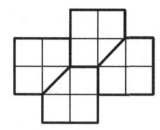

4. The heaviest is Hedgehog and the least heavy is Nyusha.

5. 30 minutes.

6. 10 rings.

7. (A) 11; (B) In the figure is shown an example:

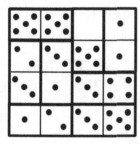

8. On the rocky island.

Olympiad 2012

1. 2012 = 1717 + 295.

2. 5 years.

3. In the figure are shown two examples:

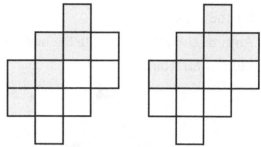

4. The filled table is shown below:

1	2	4	5	3	7	6
3	4	5	1	2	6	7
2	7	6	4	5	3	1
6	3	7	2	1	5	4
5	1	3	6	7	4	2
4	5	2	7	6	1	3
7	6	1	3	4	2	5

5. 88 *mm*.

6. 3 boxes.

7. The first grader solved three more problems than Borya.

Olympiad 2013

1. 1674 + 87 + 252 = 2013.

2. Tolya and Nikita.

3. Three times.

4. 51 minutes.

5. Anya received the piece with the number 2.

6. Here is an example:

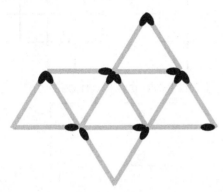

7. 20 minutes.

8. Gloria and Alex lied. Rico and Marty told the truth.

Olympiad 2014

1. His name is Nikita Andreevich.

2. 33 *km*.

3. Some options are 2 × 4 *cm* or 2.5 × 6 *cm* rectangles. Any rectangle with a side of less than 3 cm will do.

4. An example is shown in the figure:

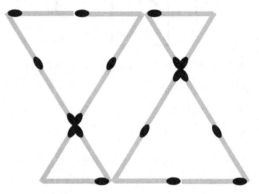

5. O+N+E+F+O+R+S+I+X = 35.

6. An example is shown below:

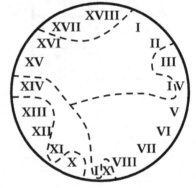

7. Popeye "the sailor" had spinach for breakfast.

8. Avoska was born on March 31.

Olympiad 2015

1. Masha - cookies, Sveta - apples, Vasya - chocolates, Petya - cake.

2. 36 solutions.

3. 5 chocolates.

4. In 1993 or in 2011.

5. An example is shown in the figure below:

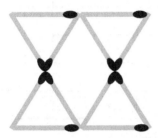

6. 12 ways.

7. He takes a handkerchief out of his pocket, puts the ball into his hat, and waves his wand.

8. There are 1 boy and 2 girls.

Olympiad 2016

1. 2 coins.

2. An example is shown below:

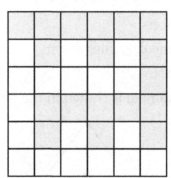

3. See the example in the figure:

4. From top to bottom: F-W-Y-L-R-P-B-K-G-O.

5. $1305 \times 6 = 7830$.

6. There are 10 cookies in a packet.

7. The prince will defeat exactly 100 dragons and the wife will be lazy.

Olympiad 2017

1. For example, 132547698 or also 153428796 and there are other options.

2. An example is shown below:

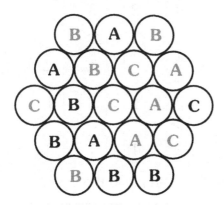

3. 3 stations.

4. Kamenyuk rotates faster.

5. 110 houses.

6. An example is shown in the figure:

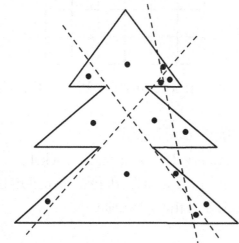

7. Valya broke the window.

8. Yes, it is possible.

Olympiad 2018

1. More expensive are 3 chocolates.

2. A possible option is: 2560 + 3564 = 6124.

3. 12 cm × 18 cm.

4. An example is shown in the figure:

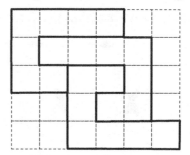

5. 11 houses.

6. 4 robots as shown in the figure below:

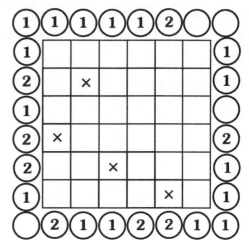

7. Every 8 minutes.

8. There are 4 liars.

Olympiad 2019

1. There were 64 ornaments on the tree.

2. Boa = 36 Parrots, Elephant = 48 Parrots, Monkey = 12 Parrots.

3. For example, $84 + 5 \times 9 - 63 : 7 = 120$ or also $97 + 5 \times 8 - 42 : 6 = 130$.

4. An example is shown in the figure:

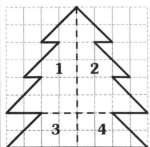 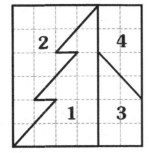

5. There were 45 children.

6. None.

7. An example is shown in the figure:

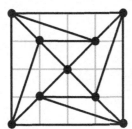

8. Yes, They will.

Olympiad 2020

1. $28 \times 9 = 252$.

2. 75 seconds.

3. $996 + 101 = 1097$.

4. It has increased by 12.

5. See the next figure:

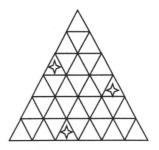

6. 2 snowballs.

7. At 12:00.

8. 1) Masha - 1°, Sasha - 3° and Olya - 5°, Kolya - 4° or also 2) Masha - 1°, Sasha - 4°, Olya - 3°, Kolya – 2°.

Answers

Made in the USA
Las Vegas, NV
06 September 2023